$1\overline{)\,1}$ $1\overline{)\,4}$ $1\overline{)\,7}$

$1\overline{)\,2}$ $1\overline{)\,5}$ $1\overline{)\,8}$

$1\overline{)\,3}$ $1\overline{)\,6}$ $1\overline{)\,9}$

$$\begin{array}{r} 2 \\ 2\overline{)\,4} \end{array}$$

$$2\overline{)\,6}$$

$$2\overline{)\,8}$$

$$2\overline{)\,10}$$

$$2\overline{)\,12}$$

$$2\overline{)\,14}$$

$$2\overline{)\,16}$$

$$2\overline{)\,18}$$

$$2\overline{)\,20}$$

WRITE ON ★ WIPE CLEAN

For use with Dry Erase Markers and Erasable Crayons

PRACTICE BOOK

Helps children learn the fundamentals of dividing numbers

Write On Wipe Clean

Laminated surface erases easily time after time!

AGES 7 & UP

DIVISION

PRACTICE POWER™

MAKES LEARNING FUN!

TO PARENTS:

Write On Wipe Clean

This booklet contains 108 division problems that your child can work out in an interactive format. The solution to each problem is listed on this page. **Practice Power**™ books are designed so children can write down their answers with dry erase marker or erasable crayon. You may check their answers after each problem or at a later time, giving you and your child flexibility. The answers may be erased for use over and over again.

DIVISION TERMS

DIVISOR
The number by which the dividend is divided.

QUOTIENT
The answer obtained when the dividend is divided by the divisor.

$$12\,\overline{)\,24}\quad = 2$$

DIVIDEND
The number or quantity to be divided.

ZEROS
When zero is divided by another number, the quotient is zero. It is not possible to divide a number by zero.

A NUMBER DIVIDED INTO ITSELF

$$12\,\overline{)\,12}\quad = 1$$

When you divide a number into itself, the answer is always one.

SHORT DIVISION
A method of dividing numbers in which much of the work is done mentally.

LONG DIVISION
A method of dividing numbers in which much of the work is written out.

$$\begin{array}{r} 12 \\ \times\ 2 \\ \hline 24 \end{array}$$

TO CHECK YOUR ANSWERS
Multiply the quotient by the divisor. The answer should be the dividend.

COUNT THE DOTS

Count the number of dots to find the quotient.

The number of dots equals the divisor.

WRITE ANSWERS HERE

$$12\,\overline{)\,24}$$

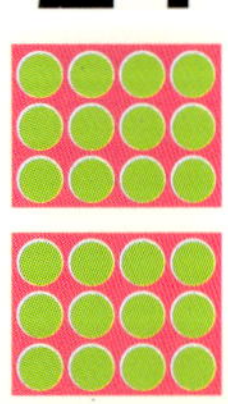

The number of dots equals the dividend.

The number of blocks equals the quotient.

DIVISION ANSWERS

1) 1 = 1	1) 2 = 2	1) 3 = 3	1) 4 = 4	1) 5 = 5	1) 6 = 6	1) 7 = 7	1) 8 = 8	1) 9 = 9
2) 4 = 2	2) 6 = 3	2) 8 = 4	2) 10 = 5	2) 12 = 6	2) 14 = 7	2) 16 = 8	2) 18 = 9	2) 20 = 10
3) 6 = 2	3) 9 = 3	3) 12 = 4	3) 15 = 5	3) 18 = 6	3) 21 = 7	3) 24 = 8	3) 27 = 9	3) 30 = 10
4) 8 = 2	4) 12 = 3	4) 16 = 4	4) 20 = 5	4) 24 = 6	4) 28 = 7	4) 32 = 8	4) 36 = 9	4) 40 = 10
5) 10 = 2	5) 15 = 3	5) 20 = 4	5) 25 = 5	5) 30 = 6	5) 35 = 7	5) 40 = 8	5) 45 = 9	5) 50 = 10
6) 12 = 2	6) 18 = 3	6) 24 = 4	6) 30 = 5	6) 36 = 6	6) 42 = 7	6) 48 = 8	6) 54 = 9	6) 60 = 10
7) 14 = 2	7) 21 = 3	7) 28 = 4	7) 35 = 5	7) 42 = 6	7) 49 = 7	7) 56 = 8	7) 63 = 9	7) 70 = 10
8) 16 = 2	8) 24 = 3	8) 32 = 4	8) 40 = 5	8) 48 = 6	8) 56 = 7	8) 64 = 8	8) 72 = 9	8) 80 = 10
9) 18 = 2	9) 27 = 3	9) 36 = 4	9) 45 = 5	9) 54 = 6	9) 63 = 7	9) 72 = 8	9) 81 = 9	9) 90 = 10
10) 20 = 2	10) 30 = 3	10) 40 = 4	10) 50 = 5	10) 60 = 6	10) 70 = 7	10) 80 = 8	10) 90 = 9	10) 100 = 10
11) 22 = 2	11) 33 = 3	11) 44 = 4	11) 55 = 5	11) 66 = 6	11) 77 = 7	11) 88 = 8	11) 99 = 9	11) 110 = 10
12) 24 = 2	12) 36 = 3	12) 48 = 4	12) 60 = 5	12) 72 = 6	12) 84 = 7	12) 96 = 8	12) 108 = 9	12) 120 = 10

PRACTICE POWER™
MAKES LEARNING FUN!

ISBN 1-930355-02-5

First published in the U.S. in 2000 by Greenbrier/Scentex.

300 Greenbrier Road, Summersville WV 26651 U.S.A.

2

3) 6

3) 15

3) 24

3) 9

3) 18

3) 27

3) 12

3) 21

3) 30

$4\overline{)\,8}$ 2

$4\overline{)\,20}$

$4\overline{)\,32}$

$4\overline{)\,12}$

$4\overline{)\,24}$

$4\overline{)\,36}$

$4\overline{)\,16}$

$4\overline{)\,28}$

$4\overline{)\,40}$

2 5 ⟌ 10	5 ⟌ 25	5 ⟌ 40
5 ⟌ 15	5 ⟌ 30	5 ⟌ 45
5 ⟌ 20	5 ⟌ 35	5 ⟌ 50

$6\overline{)12}$ 2

$6\overline{)30}$

$6\overline{)48}$

$6\overline{)18}$

$6\overline{)36}$

$6\overline{)54}$

$6\overline{)24}$

$6\overline{)42}$

$6\overline{)60}$

2

7) 14

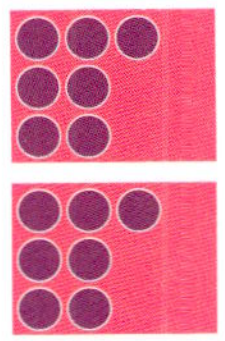

7) 35

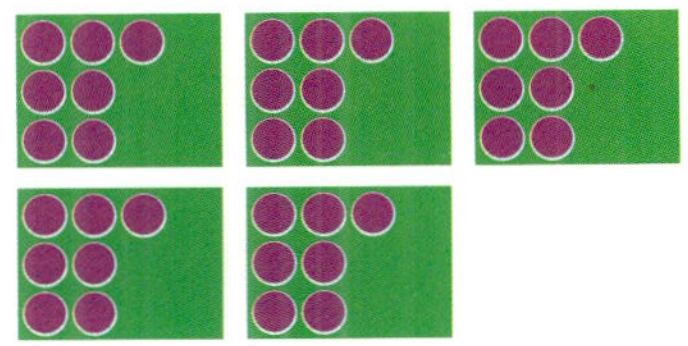

7) 56

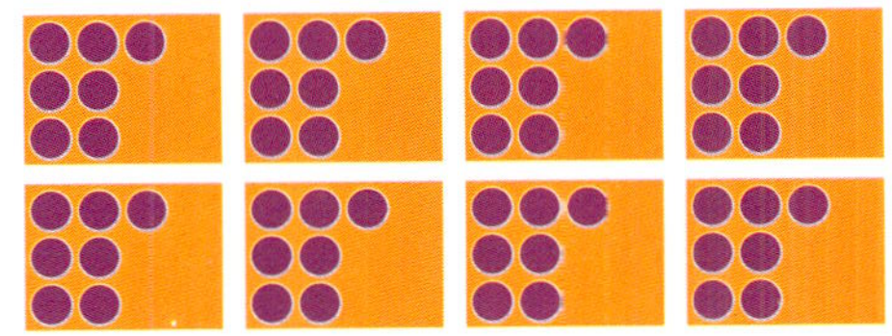

7) 21

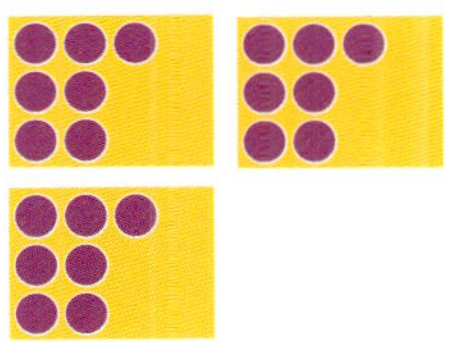

7) 42

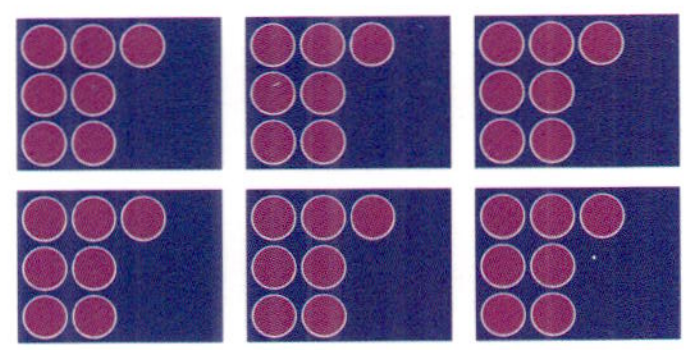

7) 63

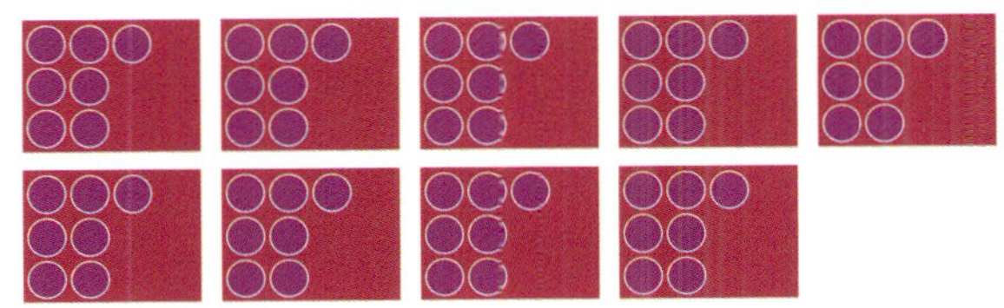

7) 28

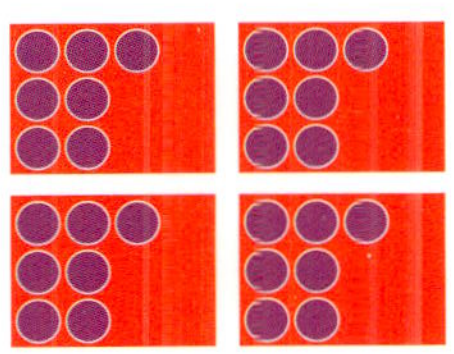

7) 49

7) 70

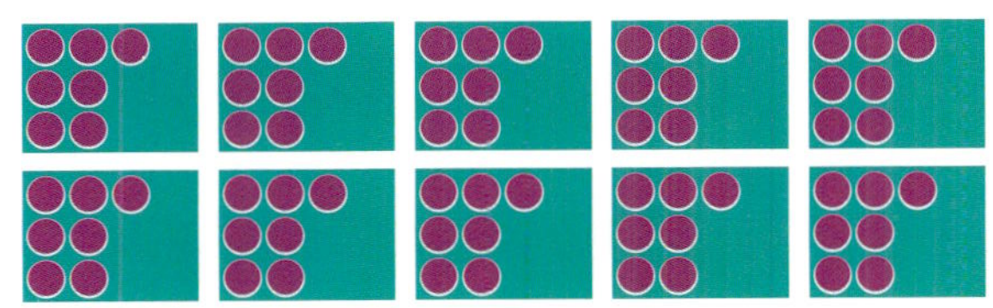

2

8) 16

8) 40

8) 64

8) 24

8) 48

8) 72

8) 32

8) 56

8) 80

2
9) 18

9) 45

9) 72

9) 27

9) 54

9) 81

9) 36

9) 63

9) 90

2

10) 20

10) 50

10) 80

10) 30

10) 60

10) 90

10) 40

10) 70

10) 100

11) 22 — 2

11) 55

11) 88

11) 33

11) 66

11) 99

11) 44

11) 77

11) 110

$12 \overline{)24}$ 2	$12 \overline{)60}$	$12 \overline{)96}$
$12 \overline{)36}$	$12 \overline{)72}$	$12 \overline{)108}$
$12 \overline{)48}$	$12 \overline{)84}$	$12 \overline{)120}$

PRACTICE POWER™

MAKES LEARNING FUN!

Make-Up And Solve Your Own Division Problems

Write On Wipe Clean

THIS BOOK BELONGS TO:

A MESSAGE TO PARENTS

The early learning years of a child's life are some of the most exciting and important he or she will ever face. It can also be a wonderful time of sharing and interaction between you and your child. We hope this practice book will serve as part of that interaction. Here are a few things you can do to help create a positive learning experience:

- ★ Have your child sit at a table in a comfortable writing position.
- ★ Use a non-toxic grease pencil, crayon or dry erase pen for best results.
- ★ Be sure your child holds the marker correctly for writing.

Most important of all, encourage your child. Practicing with this book should be a recreational activity. Let your child know it's okay (in fact, good) to make mistakes. Keeping it fun will keep them learning!

Give your child a positive learning experience. Look for other PRACTICE POWER™ educational products with our durable "Write On-Wipe Clean" surface.

ACTIVITY MATS ★ FLASH CARDS ★ LAP LEARN… PRACTICE PALS™ BOOKS ★ TRAVEL TEASERS

©GREENBRIER/SCENTEX®
300 Greenbrier Road, Summersville WV 26651 U.S.A.

PRACTICE POWER™

Write On Wipe Clean

MAKES LEARNING FUN!